Operation Code

Coding at the Water Park

Scratch 3.0

By Colleen van Lent and Kristin Fontichiaro

Cherry Lake Press

Published in the United States of America by Cherry Lake Publishing
Ann Arbor, Michigan
www.cherrylakepublishing.com

Series Adviser: Kristin Fontichiaro
Reading Adviser: Marla Conn, MS, Ed., Literacy specialist, Read-Ability, Inc.

Image Credits: Splash and backdrop images courtesy of Kristin Fontichiaro and Colleen van Lent, 4, 20; ©UnboxScience/Pixabay, 4, 20; ©Clker-Free-Vector-Images/Pixabay, 4, 20; Various images throughout courtesy of Scratch

Library of Congress Cataloging-in-Publication Data

Names: van Lent, Colleen, author. | Fontichiaro, Kristin, author.
Title: Coding at the water park / by Colleen van Lent and Kristin Fontichiaro.
Description: Ann Arbor, Michigan : Cherry Lake Publishing, 2020 | Series: Operation code | Includes index. | Audience: Grades 2-3.
Identifiers: LCCN 2019035795 (print) | LCCN 2019035796 (ebook) | ISBN 9781534159266 (hardcover) | ISBN 9781534161566 (paperback) | ISBN 9781534160415 (pdf) | ISBN 9781534162716 (ebook)
Subjects: LCSH: Scratch (Computer program language)–Juvenile literature. | Computer programming–Juvenile literature. | Swimming–Computer simulation–Juvenile literature.
Classification: LCC QA76.73.S345 V357 2020 (print) | LCC QA76.73.S345 (ebook) | DDC 005.13/3–dc23
LC record available at https://lccn.loc.gov/2019035795
LC ebook record available at https://lccn.loc.gov/2019035796

Cherry Lake Publishing would like to acknowledge the work of the Partnership for 21st Century Learning, a Network of Battelle for Kids. Please visit *http://www.battelleforkids.org/networks/p21* for more information.

Printed in the United States of America
Corporate Graphics

NOTE TO READERS: Use this book to practice your Scratch 3 coding skills. If you have never used Scratch before, ask a parent, teacher, or librarian to help you set up an account at *https://scratch.mit.edu*. Read the tutorials on the website to learn how Scratch works. Then you will be ready for the activities in this book! You will practice using variables, if/then statements, copying code to other sprites, using effects to change a sprite's look, and more! Find all the starter and final programs at *https://scratch.mit.edu/users/CherryLakeCoding*.

Table of Contents

I'm ready!
You've got this!

Make a Splash!

Lucas has always wanted to dive off the diving board. I brought him to the pool to practice. Let's use Scratch to help him make the perfect dive!

I got things started at *https://scratch.mit.edu/projects/315014087*.

Can you help me with the rest?

Pro Tip!

Scratch lets you see other people's projects and copy them to make them your own. My code contains the **sprites**, sound, and backdrops we will use in this book.

Ways to Make a Sprite Move

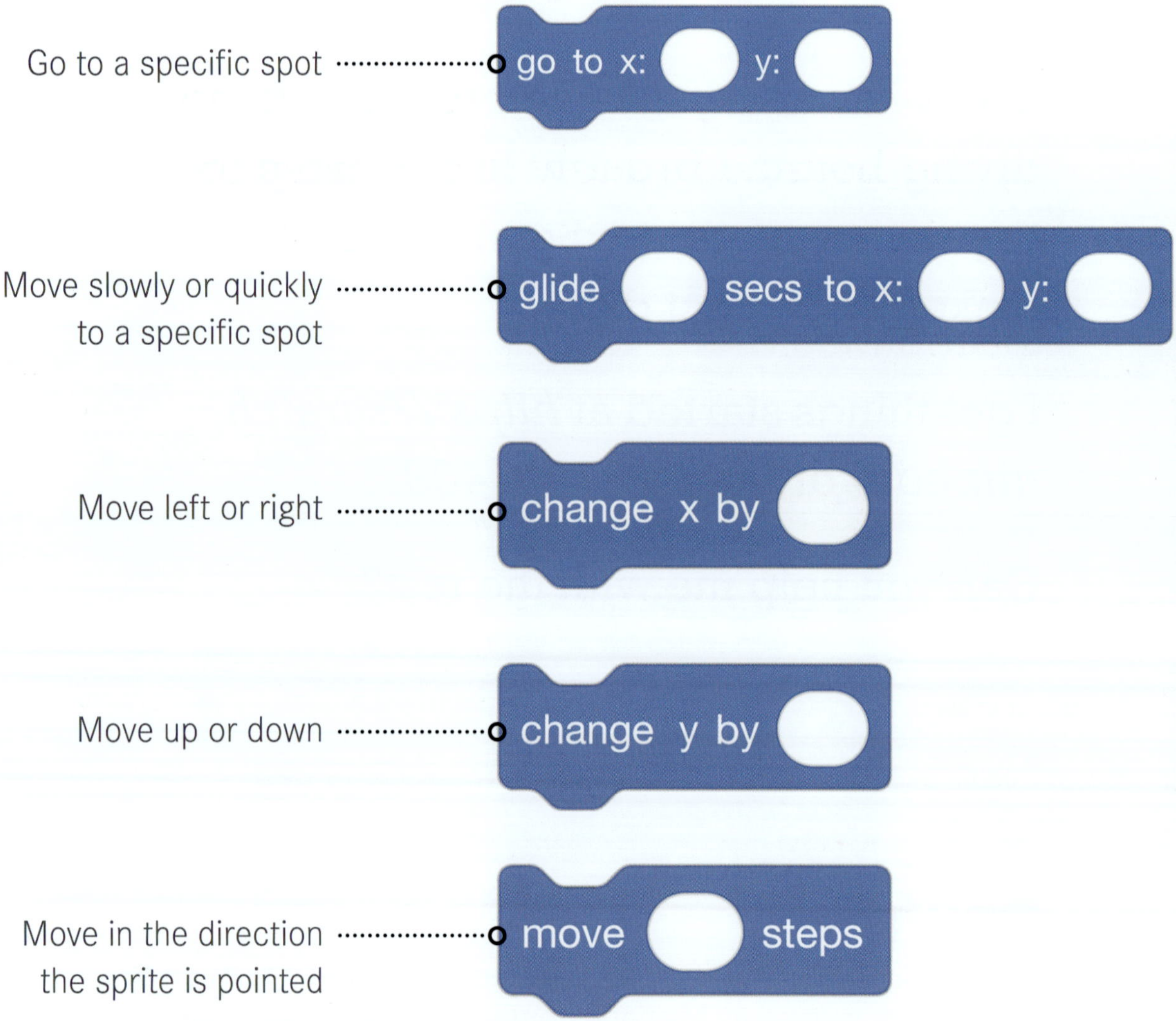

Motion Blocks

There are many different ways to move a sprite. Let's code a few to get Lucas onto the diving board and to the edge.

Here's the code I made:

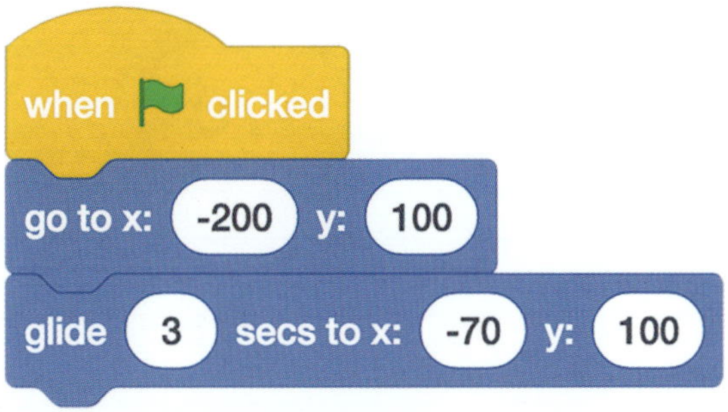

Pro Tip!

The *go to, glide,* and *move* **Motion** blocks can move left, right, or diagonally. The *change x* and *change y* **Motion** blocks can only move in one direction.

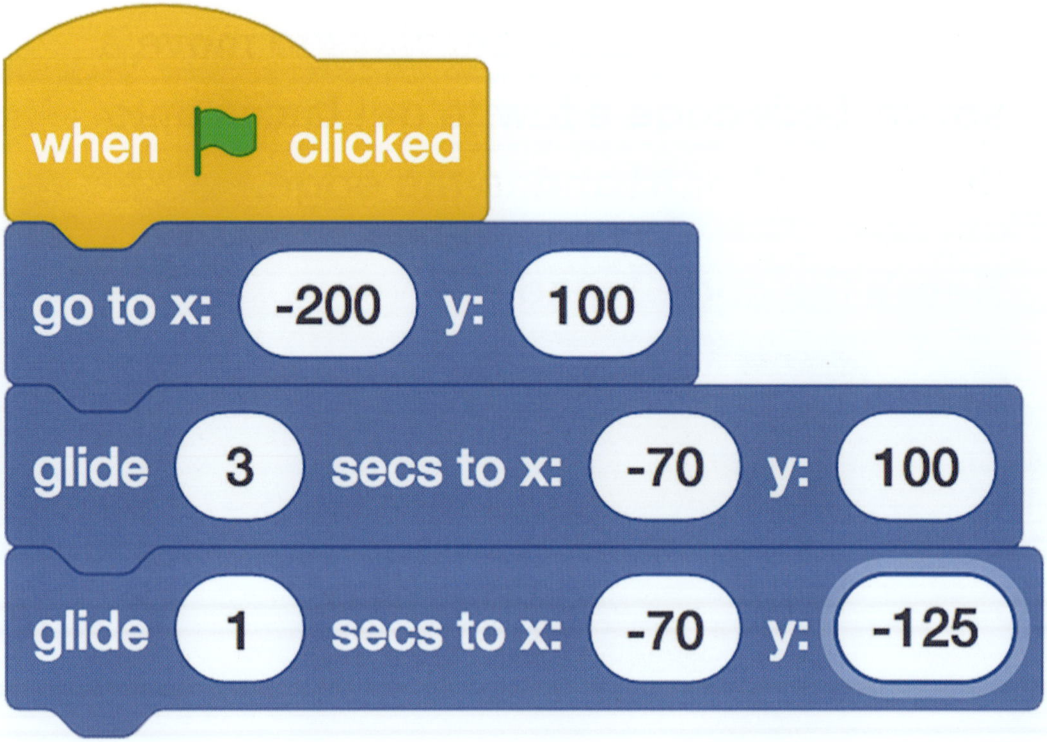
when clicked
go to x: -200 y: 100
glide 3 secs to x: -70 y: 100
glide 1 secs to x: -70 y: -125

First Dive

Now it's time to dive. Which block should we use? One option is to use *glide* again. Let's see what happens when we add this *glide* Motion block.

```
glide (1) secs to x: (-70) y: (-125)
```

Test your code. What happens?

Pro Tip!

The bigger the first number in the *glide* block is, the slower the sprite will move.

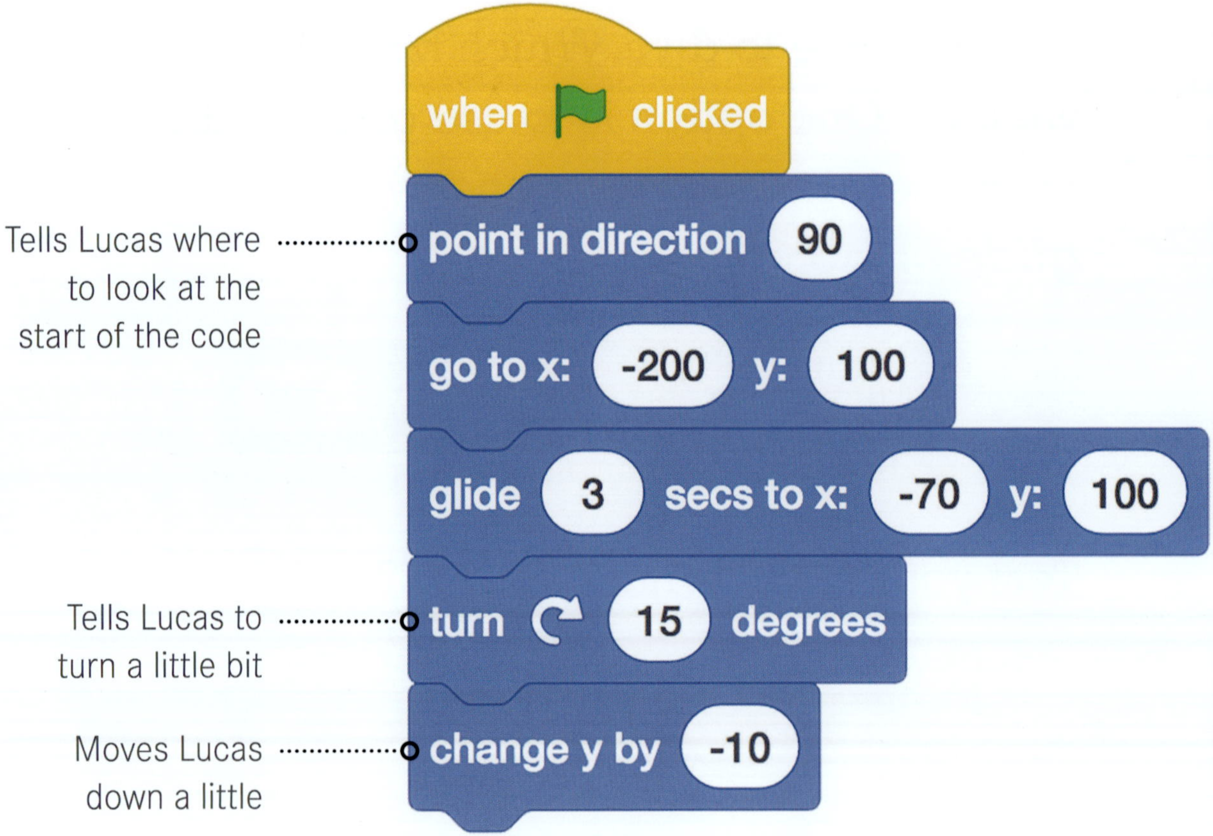

Tells Lucas where to look at the start of the code
when clicked
point in direction 90
go to x: -200 y: 100
glide 3 secs to x: -70 y: 100
Tells Lucas to turn a little bit
turn 15 degrees
Moves Lucas down a little
change y by -10

Coding a Better Dive

That second *glide* block didn't look as good as we'd hoped. We want Lucas to dive, not fall! Click and drag `glide (1) secs to x: (-70) y: (-125)` away to delete it.

Let's turn him in the air as he goes down. We'll have him turn a little bit, then go down.

Add these three new blocks of code so it looks like ours. Test your code. What happened?

Pro Tip!

When using a *turn* **Motion** block, you may need to add code to reset the sprite upright when you run the program again. We use the *point in direction* **Motion** block to do that. If you don't have that block, Lucas starts off tilted the next time you run the program!

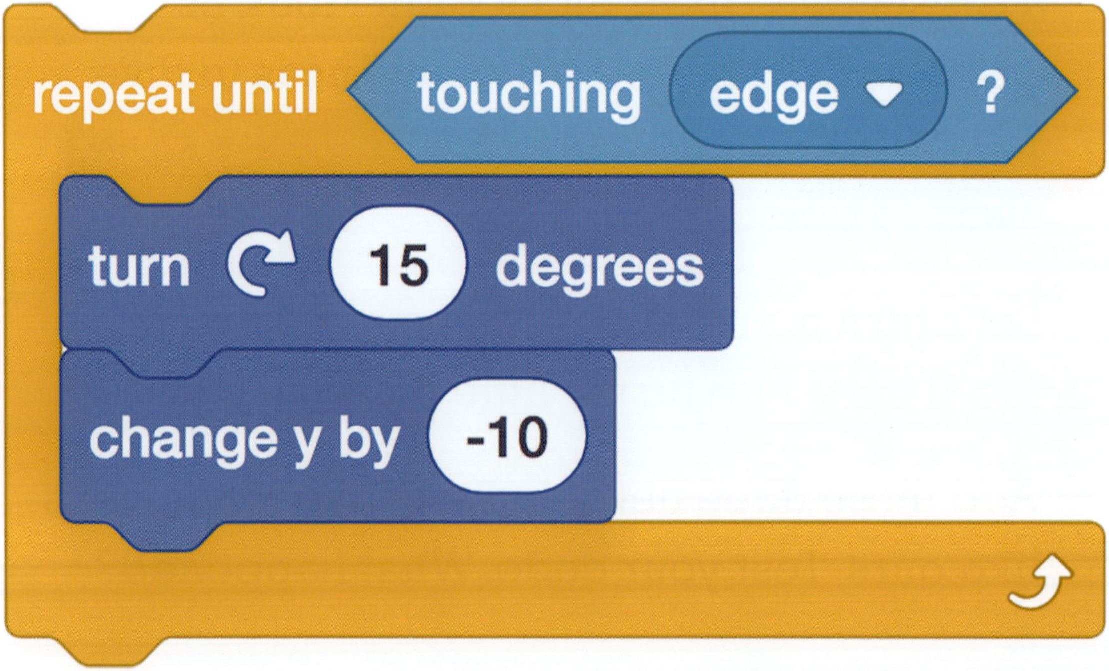
repeat until
touching
edge
?
turn
15
degrees
change y by
-10

Turn and Turn

A good dive needs many turns. Let's code this!

Our first step is to put this Control block around the *turn* and *change y* Motion blocks:

Add the code. Test the program. What do you notice?

Pro Tip!

Sometimes when you drag a *repeat* block, it doesn't end up where you were hoping. Don't worry—it even happens to Scratch pros! One trick is to drag the *repeat* block right above the first block you want to repeat.

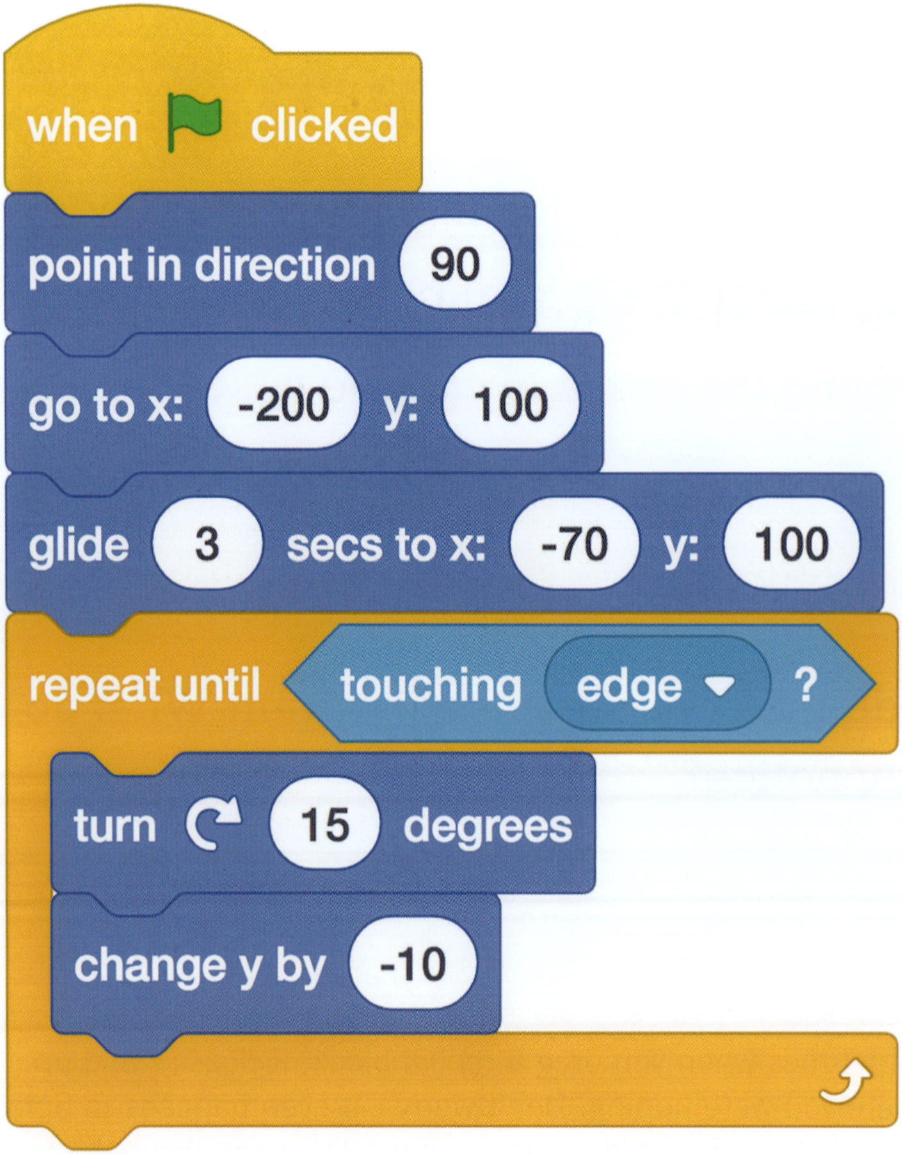
when clicked
point in direction 90
go to x: -200 y: 100
glide 3 secs to x: -70 y: 100
repeat until touching edge ?
turn 15 degrees
change y by -10

Touching Edge

Lucas looks goofy spinning in the water! We need to tell the *repeat* block when to stop.

One way to do this is with Sensing blocks.

Drag this Sensing block [touching edge ?] into your *repeat until* Control block.

Edge here means the end of the Stage. This stops Lucas's turn-and-fall motion when his sprite touches the bottom.

Test your code.

Pro Tip!

Sometimes, there is more than one way to make things happen in Scratch. How could you use *touching color* instead?

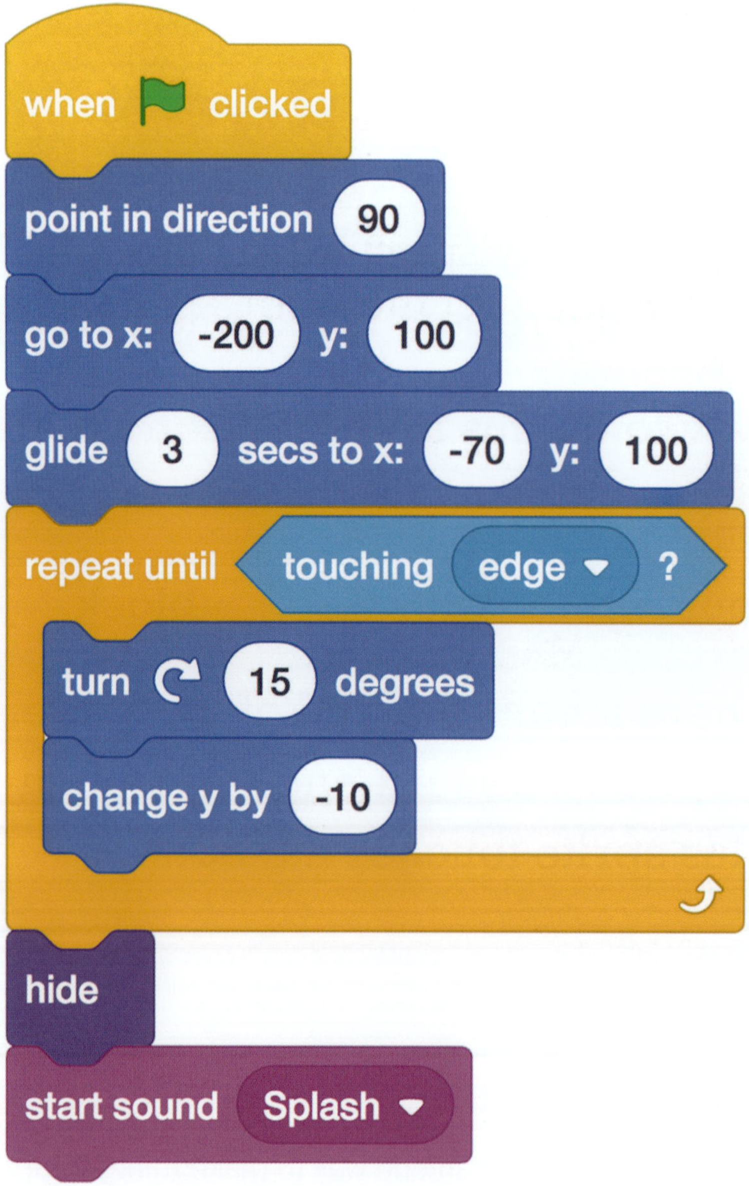
when clicked
point in direction 90
go to x: -200 y: 100
glide 3 secs to x: -70 y: 100
repeat until touching edge ?
turn 15 degrees
change y by -10
hide
start sound Splash

Hide and Splash

Success! Almost. Lucas shouldn't hang out at the bottom of the pool. Let's have Lucas disappear into the water using the hide Looks block.

And for fun, let's add this Sound block as well:

start sound Splash ▾

Test your code at least twice. Oops! What happened?

Pro Tip!

Lucas didn't have a **default** sound. We had to open the Sounds tab (next to Costumes) and select a new sound.

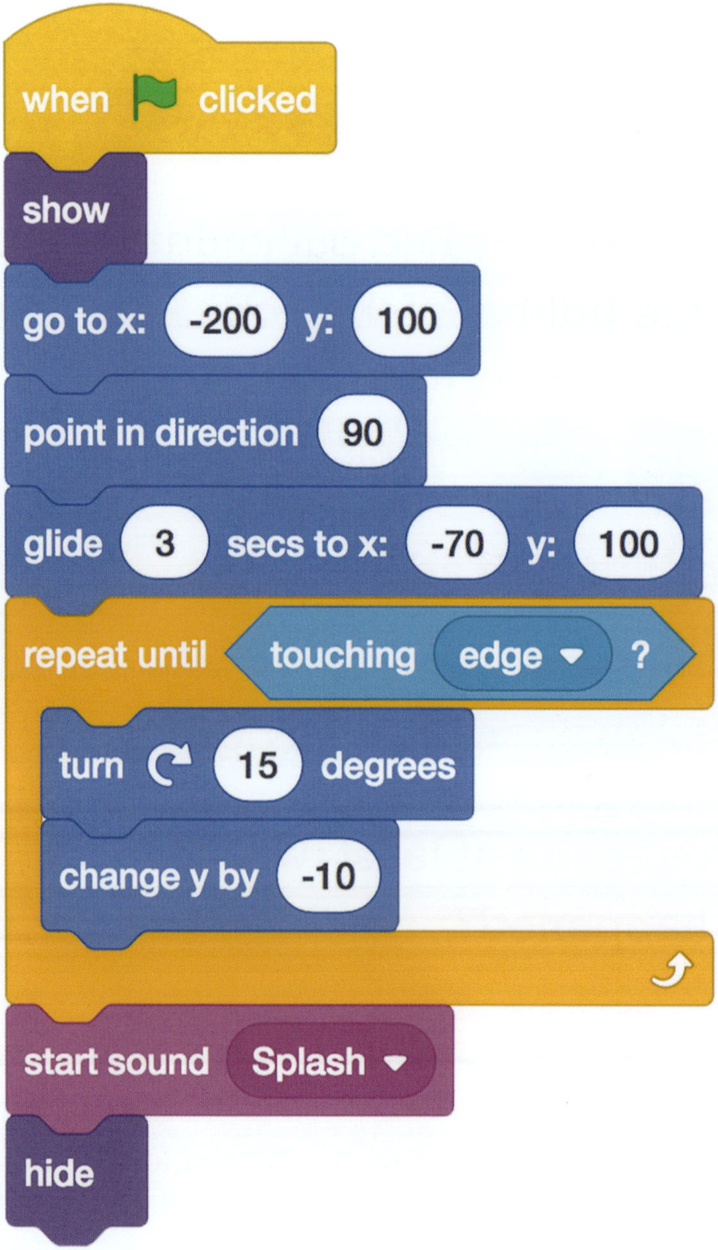
when clicked
show
go to x: -200 y: 100
point in direction 90
glide 3 secs to x: -70 y: 100
repeat until touching edge ?
turn 15 degrees
change y by -10
start sound Splash
hide

Where's Lucas?

Good thing we tested our code again. Lucas disappeared! That is why it's always a good idea to test your code twice in Scratch.

We can get Lucas back by adding the show block at the start of the program. You can find this block in the Looks **category**.

Remember, finding and fixing mistakes is normal in coding. There's even a name for it: **debugging**.

Pro Tip!

Don't wait until your code is done to test it. Good coders test their code in small chunks. This makes it easier to find mistakes.

Way to Go!

Thank you for helping Lucas nail that dive! What else can you help him do?

Here are a few ideas:

- Change the Splash backdrop when Lucas touches the edge. Remember to change back to the Pool backdrop at the start of your code!
- Make his dive faster or slower.
- Have him bounce before he dives.

What other ideas do you have?

Pro Tip!

If you would like to see our final code, go to *https://scratch.mit.edu/projects/315072058*.

Glossary

category (KAT-uh-gor-ee) a group of things that have something in common

debugging (dee-BUG-ing) finding and fixing mistakes in your code

default (DEE-fawlt) the choice Scratch makes for you in advance that you can change

sprites (SPRYTS) characters or objects in Scratch

Find Out More

Books

LEAD Project. *Super Scratch Programming Adventure!* San Francisco, California: No Starch Press, 2019.

Lovett, Amber. *Coding with Blockly.* Ann Arbor, Michigan: Cherry Lake Publishing, 2017.

Websites

Scratch
http://scratch.mit.edu
Build your Scratch code online at this site.

Scratch Wiki
https://en.scratch-wiki.info
If you get stuck, ask an adult to help you look on this site for advice.

Index

About the Authors

Colleen van Lent teaches coding and Web design at the University of Michigan School of Information. She has three cool kids and a dog named Bacon. She wishes she could touch her toes.

Kristin Fontichiaro teaches at the University of Michigan School of Information. She likes working with kids on creative projects from coding to sewing to junk box inventions. She has written or edited almost 100 books for kids.